THE EVOLUTION OF INNOVATION:
100 INVENTIONS THAT CHANGED THE WORLD

The Stories Behind the Inventions That Transformed Life as We Know It

DESCRIPTION:

The Evolution of Innovation: 100 Inventions That Changed the World takes you on an insightful journey through the most transformative inventions in human history. From the invention of the wheel to the dawn of artificial intelligence, this book explores the remarkable stories behind the technologies and ideas that have shaped our everyday lives.

Each invention is brought to life with detailed accounts of its origins, the brilliant minds behind it, and its far-reaching impact on society. You'll discover how groundbreaking innovations in science, medicine, engineering, and communication have fueled human progress—from ancient tools that revolutionized agriculture and transportation, to modern marvels like computers, smartphones, and renewable energy.

Whether it's the printing press that spread knowledge across the globe, the steam engine that powered the Industrial Revolution, or the internet that transformed the way we connect, this book offers a deep dive into the fascinating world of invention and discovery. Perfect for history buffs, tech enthusiasts, or anyone curious about the forces that have shaped civilization, The Evolution of Innovation reveals how human ingenuity continues to drive us forward.

TABLE OF CONTENTS:

Chapter 1: Ancient Inventions

1. The Wheel (c. 3500 BCE)
2. Writing (c. 3200 BCE)
3. The Plow (c. 3000 BCE)
4. Papyrus (c. 3000 BCE)
5. Sailing Ships (c. 3000 BCE)
6. Concrete (c. 1300 BCE)
7. The Compass (c. 206 BCE)
8. The Abacus (c. 2000 BCE)
9. The Calendar (c. 3000 BCE)
10. Aqueducts (c. 312 BCE)

Chapter 2: Medieval Inventions

11. The Printing Press (1440 CE)
12. Gunpowder (9th century CE)
13. The Mechanical Clock (13th century CE)
14. Eyeglasses (13th century CE)
15. The Spinning Wheel (11th century CE)
16. The Windmill (9th century CE)
17. Paper Money (11th century CE)
18. The Astrolabe (c. 150 BCE, refined in the 8th century CE)
19. Distillation (8th century CE)
20. Arabic Numerals (9th century CE)

Chapter 3: Renaissance and Early Modern Inventions

21. The Telescope (1608 CE)
22. The Microscope (1590 CE)
23. The Flush Toilet (1596 CE)
24. The Steam Engine (1698 CE)
25. The Barometer (1643 CE)
26. The Thermometer (1612 CE)
27. The Printing Press with Movable Type (1440 CE)
28. The Mechanical Loom (1801 CE)
29. Vaccination (1796 CE)
30. The Hot Air Balloon (1783 CE)

Chapter 4: 19th-Century Inventions

31. The Telegraph (1837 CE)
32. The Light Bulb (1879 CE)
33. The Telephone (1876 CE)
34. The Internal Combustion Engine (1859 CE)
35. The Typewriter (1868 CE)
36. Photography (1826 CE)
37. The Sewing Machine (1846 CE)
38. Dynamite (1867 CE)
39. The Bicycle (1817 CE)
40. The Pasteurization Process (1864 CE)

Chapter 5: Early 20th Century Inventions

41. The Airplane (1903 CE)
42. The Model T Ford (1908 CE)
43. Penicillin (1928 CE)
44. The Radio (1901 CE)
45. The Vacuum Cleaner (1901 CE)
46. Plastic (1907 CE)
47. The X-ray Machine (1895 CE)
48. The Helicopter (1936 CE)
49. The Refrigerator (1913 CE)
50. The Neon Light (1910 CE)

Chapter 6: Mid-20th Century Inventions

51. The Jet Engine (1930s CE)
52. Television (1927 CE)
53. Nuclear Power (1940s CE)
54. The Computer (1940s CE)
55. The Microwave Oven (1945 CE)
56. Velcro (1941 CE)
57. The Transistor (1947 CE)
58. The Credit Card (1950 CE)
59. The Hovercraft (1956 CE)
60. The Polio Vaccine (1955 CE)

Chapter 7: Late 20th Century Inventions

61. The Internet (1960s-1990s CE)
62. The Personal Computer (1970s CE)
63. The Mobile Phone (1973 CE)
64. The Artificial Heart (1982 CE)
65. GPS (1973 CE)
66. The Space Shuttle (1981 CE)
67. DNA Fingerprinting (1984 CE)
68. The ATM (1967 CE)
69. Post-It Notes (1974 CE)
70. The MRI Scanner (1977 CE)

Chapter 8: Early 21st Century Inventions

71. The Hybrid Car (1997 CE)
72. Social Media Platforms (2000s CE)
73. The iPhone (2007 CE)
74. 3D Printing (1980s-2000s CE)
75. Bitcoin (2009 CE)
76. CRISPR Gene Editing (2012 CE)
77. The Electric Car (2008 CE)
78. Cloud Computing (2000s CE)
79. Artificial Intelligence Assistants (2010s CE)
80. Wearable Fitness Trackers (2010s CE)

Chapter 9: Emerging Innovations

81. Self-Driving Cars (2010s CE)
82. 5G Technology (2020s CE)
83. Lab-Grown Meat (2020s CE)
84. Quantum Computing (2020s CE)
85. Vertical Farming (2010s CE)
86. Exoskeletons (2020s CE)
87. Hyperloop (2010s CE)
88. Augmented Reality (2010s CE)
89. Holographic Displays (2020s CE)
90. Biodegradable Plastics (2020s CE)

Chapter 10: Future Inventions

91. Artificial General Intelligence (AGI)
92. Fusion Energy
93. Smart Cities
94. Brain-Computer Interfaces (BCIs)
95. Teleportation
96. Space Tourism
97. Artificial Organs
98. Nanomedicine
99. Flying Cars
100. Interstellar Travel

CHAPTER 1: ANCIENT INVENTIONS

1. THE WHEEL (C. 3500 BCE)

Description: One of the most transformative inventions, the wheel enabled easier transportation of goods and people. Initially used in pottery-making, its true potential was realized in Mesopotamia for use in carts and chariots.

History: The first wheels were made from wooden planks bound together. Their use revolutionized agriculture, transport, and trade across early civilizations like Mesopotamia and Egypt.

2. WRITING (C. 3200 BCE)

Description: Writing allowed civilizations to record laws, stories, trade, and history. The earliest known form of writing is cuneiform, developed by the Sumerians.

History: Writing began as a series of pictographs but evolved into systems like cuneiform in Mesopotamia and hieroglyphs in Egypt. It was key to the development of complex societies.

3. THE PLOW (C. 3000 BCE)

Description: The plow enabled humans to till the soil more efficiently, supporting the rise of agricultural societies. Early plows were simple wooden tools pulled by oxen.

History: The first plows appeared in Mesopotamia and Egypt. The shift from simple hand tools to plows allowed civilizations to grow enough food to sustain larger populations.

4. PAPYRUS (C. 3000 BCE)

Description: Papyrus, a precursor to paper, was used by ancient Egyptians to create scrolls. It was made from the pith of the papyrus plant.

History: Egyptians used papyrus for record-keeping and religious texts. Its lightweight and portability made it an essential tool for bureaucracy and knowledge transmission.

5. SAILING SHIPS (C. 3000 BCE)

Description: Early sailing ships allowed civilizations to navigate rivers and seas, expanding trade and exploration. Egyptians were among the first to build ships with sails.

History: Initially used on the Nile, these vessels expanded trade routes and military power, eventually influencing the design of larger ships in Phoenicia and Greece.

6. CONCRETE (C. 1300 BCE)

Description: The Romans developed concrete, a mixture of volcanic ash, lime, and water. Its strength and durability allowed for monumental architecture, like the Pantheon.

History: Roman concrete was more durable than modern concrete due to its unique ingredients, particularly volcanic ash. This material enabled the Romans to build aqueducts, roads, and temples that have lasted for centuries.

7. THE COMPASS (C. 206 BCE)

Description: Invented during the Han Dynasty in China, the compass revolutionized navigation by allowing sailors to determine direction even on cloudy days.

History: The earliest compasses were used for divination but soon found use in navigation. Chinese and Arab traders spread the compass, which reached Europe by the 12th century, becoming a vital tool for explorers.

8. THE ABACUS (C. 2000 BCE)

Description: One of the earliest calculators, the abacus helped traders and merchants keep track of large numbers.

History: The abacus likely originated in Mesopotamia and was refined in China. It remained in use for centuries as a practical tool for quick calculations in markets and businesses.

9. THE CALENDAR (C. 3000 BCE)

Description: Early calendars were created to predict natural events like the flooding of the Nile. The Egyptian calendar, based on lunar and solar cycles, is one of the earliest known.

History: Ancient civilizations like the Egyptians, Mayans, and Babylonians developed calendars to track time, seasons, and agricultural cycles, laying the foundation for modern calendars.

10. AQUEDUCTS (C. 312 BCE)

Description: The Romans developed aqueducts to transport water from distant sources into cities, supporting large urban populations.

History: Roman engineers built aqueducts using stone, concrete, and arches to channel water over long distances. This innovation allowed Roman cities to thrive by providing a steady supply of fresh water.

CHAPTER 2: MEDIEVAL INVENTIONS
11. THE PRINTING PRESS (1440 CE)

Description: Johannes Gutenberg's invention of the printing press revolutionized the production of books and the dissemination of knowledge, particularly in Europe.

History: Gutenberg's press used movable type, allowing for faster, cheaper printing. It played a key role in spreading the Reformation, Renaissance ideas, and scientific knowledge.

12. GUNPOWDER (9TH CENTURY CE)

Description: Invented in China, gunpowder was initially used for fireworks but later became a critical component in warfare, leading to the development of cannons, guns, and explosives.

History: Gunpowder technology spread to the Islamic world and Europe, changing the nature of warfare by making traditional armor and fortifications less effective.

13. THE MECHANICAL CLOCK (13TH CENTURY CE)

Description: Mechanical clocks were developed to track time more accurately than sundials or water clocks. The first clocks appeared in monasteries to regulate prayers.

History: The invention of the escapement mechanism allowed for the precise measurement of time. Clocks spread to cities, becoming essential for regulating commerce and daily life.

14. EYEGLASSES (13TH CENTURY CE)

Description: The invention of eyeglasses in Italy revolutionized the lives of scholars and artisans by allowing those with poor eyesight to see clearly.

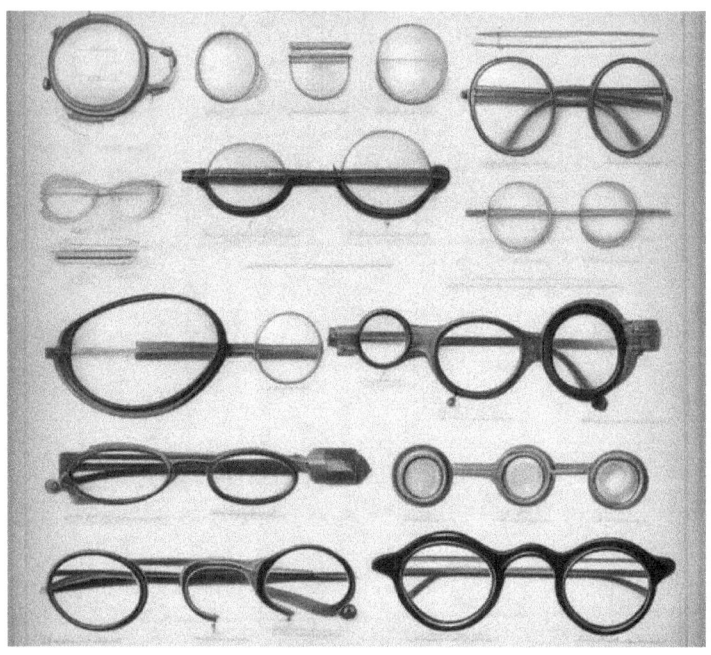

History: Early eyeglasses had convex lenses for farsightedness. Over time, they became more accessible and were refined to include lenses for nearsightedness as well.

15. THE SPINNING WHEEL (11TH CENTURY CE)

Description: Originating in India, the spinning wheel sped up the process of turning fiber into yarn, revolutionizing the textile industry.

History: The spinning wheel became widespread in Europe and Asia, dramatically increasing textile production and paving the way for later industrial innovations in weaving and clothing.

16. THE WINDMILL (9TH CENTURY CE)

Description: Windmills harnessed wind power to grind grain, pump water, and perform other tasks. Early windmills appeared in Persia and spread to Europe.

History: Windmills were critical for agriculture in regions with few rivers for watermills. By the 12th century, they were common in Europe, playing a major role in food production.

17. PAPER MONEY (11TH CENTURY CE)

Description: First developed in China during the Song Dynasty, paper money made commerce more efficient by replacing bulky metal coins.

History: The innovation spread to the Islamic world and later Europe. Paper money allowed for easier trade and the expansion of financial systems, eventually evolving into modern currency.

18. THE ASTROLABE
(C. 150 BCE, REFINED IN THE 8TH CENTURY CE)

Description: An ancient tool for measuring the altitude of celestial bodies, the astrolabe was refined by Islamic scholars and became crucial for navigation and astronomy.

History: Astrolabes allowed sailors to determine their latitude at sea, aiding exploration and trade. The device also played a key role in the development of astronomy in the Islamic Golden Age.

19. DISTILLATION (8TH CENTURY CE)

Description: Arab alchemists refined the process of distillation, which allowed for the purification of liquids like alcohol and essential oils, as well as the development of stronger medicines.

History: The technique spread to Europe, where it was used in the production of spirits, perfumes, and medicinal compounds, contributing to advances in chemistry and medicine.

20. ARABIC NUMERALS (9TH CENTURY CE)

Description: Based on Indian mathematics and introduced to Europe by Arab scholars, Arabic numerals, including the concept of zero, revolutionized mathematics and commerce.

History: Arabic numerals replaced Roman numerals due to their simplicity in calculations, particularly for trade and science, helping advance algebra, geometry, and engineering.

CHAPTER 3: RENAISSANCE AND EARLY MODERN INVENTIONS
21. THE TELESCOPE (1608 CE)

Description: The invention of the telescope by Dutch optician Hans Lippershey, and its refinement by Galileo, transformed astronomy by allowing humans to see far beyond Earth.

History: Galileo's observations of moons around Jupiter and the phases of Venus challenged the geocentric model of the universe, setting the stage for modern astronomy.

22. THE MICROSCOPE (1590 CE)

Description: Developed by Dutch spectacle-makers Hans and Zacharias Janssen, the microscope opened up a previously invisible world of microorganisms and cells.

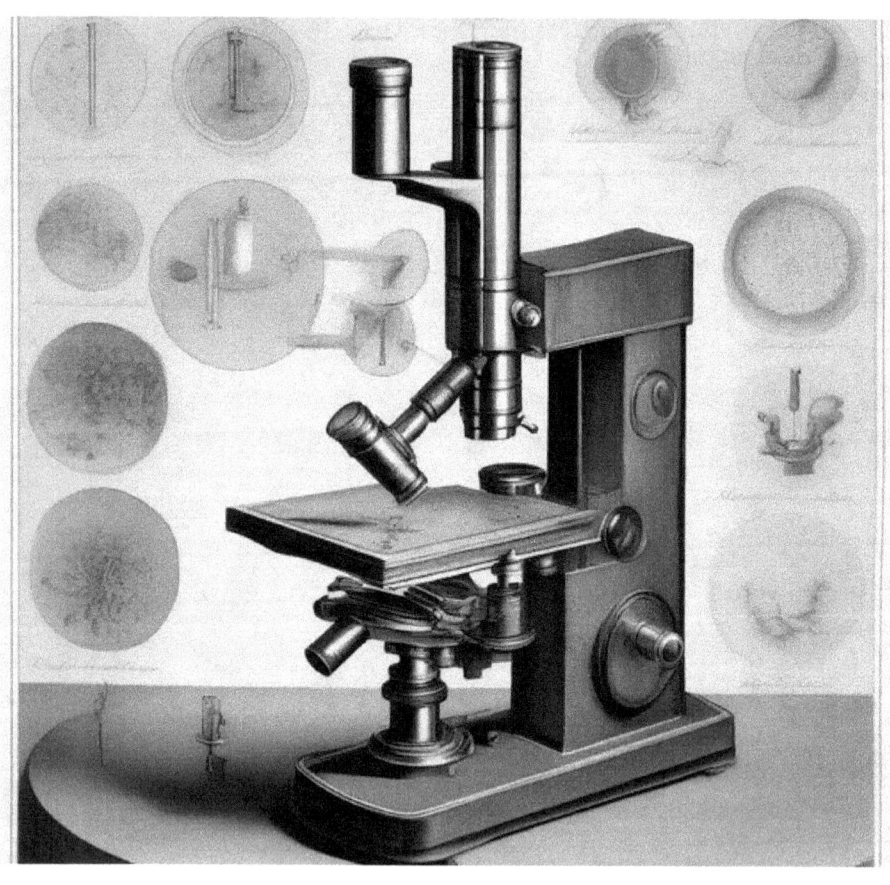

History: Robert Hooke's 1665 book Micrographia documented observations of the microscopic world, leading to advancements in biology and medicine.

23. THE FLUSH TOILET (1596 CE)

Description: Sir John Harington invented the first flush toilet in England. It consisted of a cistern that released water to flush waste, improving sanitation.

History: Although initially considered a luxury, Harington's invention laid the foundation for modern sanitation systems, reducing disease in urban areas.

24. THE STEAM ENGINE (1698 CE)

Description: Thomas Savery developed the first practical steam engine, used primarily to pump water out of mines. James Watt's later improvements made it more efficient.

History: Watt's steam engine powered the Industrial Revolution, transforming industries such as textiles, mining, and transportation by providing a reliable source of mechanical power.

25. THE BAROMETER (1643 CE)

Description: Evangelista Torricelli invented the barometer to measure atmospheric pressure, leading to better weather prediction and an understanding of meteorology.

History: The barometer became a key instrument in scientific experiments and practical weather forecasting, helping sailors and farmers alike.

26. THE THERMOMETER (1612 CE)

Description: Galileo Galilei developed the first thermoscope, a precursor to the modern thermometer. It allowed for the measurement of temperature changes.

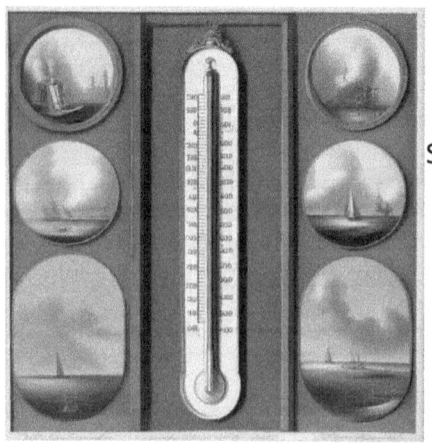

History: The thermometer was further refined by scientists like Daniel Gabriel Fahrenheit and Anders Celsius, who created temperature scales still in use today.

27. THE PRINTING PRESS WITH MOVABLE TYPE (1440 CE)

Description: Johannes Gutenberg's movable type press allowed for the mass production of books and other written materials, significantly lowering costs and increasing accessibility.

History: The printing press played a major role in the spread of ideas during the Renaissance and Reformation, and it remains one of the most important inventions in the history of communication.

28. THE MECHANICAL LOOM (1801 CE)

Description: Joseph Marie Jacquard developed the mechanical loom, which used punched cards to control weaving patterns, a precursor to programmable machines.

History: The Jacquard loom revolutionized textile manufacturing and directly influenced the development of early computers, thanks to its use of binary-like punch cards.

29. VACCINATION (1796 CE)

Description: Edward Jenner discovered that exposure to cowpox could protect against smallpox, leading to the first vaccine and the birth of modern immunology.

History: Jenner's work eventually led to the global eradication of smallpox and inspired the development of vaccines for other diseases, saving millions of lives.

30. THE HOT AIR BALLOON (1783 CE)

Description: The Montgolfier brothers created the first successful hot air balloon, allowing humans to experience controlled flight for the first time.

History: Their early flights in France sparked public fascination and laid the groundwork for modern aviation, which would take further leaps with the invention of the airplane.

CHAPTER 4: 19TH-CENTURY INVENTIONS

31. THE TELEGRAPH (1837 CE)

Description: Samuel Morse developed the telegraph, a device that allowed for the instant transmission of messages over long distances using electrical signals.

History: The telegraph revolutionized communication, particularly for business and military use. It laid the foundation for the modern global communications network.

32. THE LIGHT BULB (1879 CE)

Description: Thomas Edison and Joseph Swan both developed practical incandescent light bulbs that provided reliable indoor lighting.

History: Edison's improvements to the light bulb, along with his system for generating and distributing electricity, transformed daily life by extending productive hours and reducing reliance on fire-based lighting.

33. THE TELEPHONE (1876 CE)

Description: Alexander Graham Bell invented the first practical telephone, allowing for real-time voice communication over long distances.

History: The telephone quickly became an essential tool for business and personal communication, shrinking the world by making distant voices instantly accessible.

34. THE INTERNAL COMBUSTION ENGINE (1859 CE)

Description: Étienne Lenoir developed the first internal combustion engine, which used the explosion of fuel to produce power.

History: The internal combustion engine paved the way for the automobile, revolutionizing transportation and industry by making personal and freight transportation faster and more flexible.

35. THE TYPEWRITER (1868 CE)

Description: Christopher Sholes invented the typewriter, a machine that allowed for faster and more efficient writing.

History: The typewriter became a staple of offices and businesses, transforming the way documents were produced and eventually leading to the development of computers and word processors.

36. PHOTOGRAPHY (1826 CE)

Description: Nicéphore Niépce created the first photograph, a rudimentary process that involved exposing a bitumen-coated plate to light.

History: Photography evolved quickly, with Louis Daguerre's daguerreotype and subsequent improvements making it more practical. It transformed journalism, art, and personal memory-keeping.

37. THE SEWING MACHINE (1846 CE)

Description: Elias Howe invented the first successful sewing machine, which mechanized the process of stitching fabric.

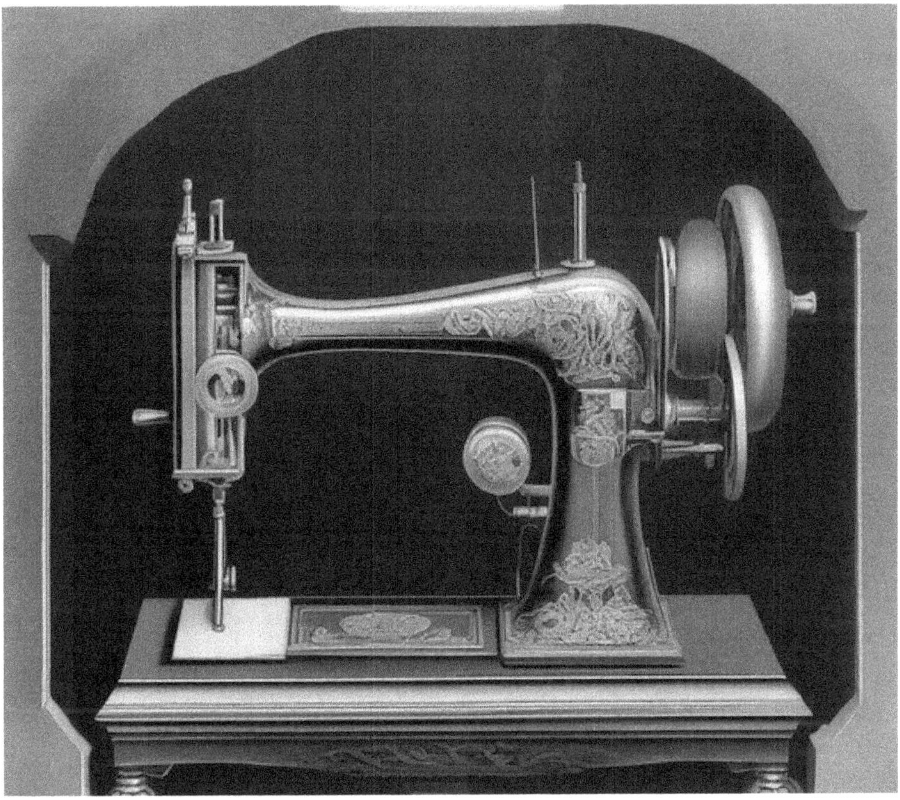

History: The sewing machine revolutionized clothing production, drastically reducing the time needed to make garments and contributing to the growth of the textile industry.

38. DYNAMITE (1867 CE)

Description: Alfred Nobel invented dynamite, a powerful explosive that was safer to handle than earlier forms of nitroglycerin.

History: Dynamite was used in construction and mining but also in warfare, much to Nobel's regret. His invention earned him great wealth, which he used to establish the Nobel Prizes.

39. THE BICYCLE (1817 CE)

Description: Karl Drais invented the first bicycle-like machine, known as the "running machine" or "draisine," which had no pedals but was propelled by pushing off the ground with the feet.

History: Over the next few decades, bicycles evolved into their modern form, providing an affordable, eco-friendly mode of transport that became popular worldwide.

40. THE PASTEURIZATION PROCESS (1864 CE)

Description: Louis Pasteur developed pasteurization, a process that involves heating liquids to kill harmful bacteria.

History: Pasteur's invention was initially used for wine and beer but soon extended to milk, drastically improving food safety and public health by reducing foodborne illnesses.

CHAPTER 5: EARLY 20TH CENTURY INVENTIONS

41. THE AIRPLANE (1903 CE)

Description: The Wright brothers, Orville and Wilbur, achieved the first powered, controlled flight with their airplane at Kitty Hawk, North Carolina.

History: Their success marked the beginning of modern aviation, which would evolve into commercial and military air travel, transforming global transportation and warfare.

42. THE MODEL T FORD (1908 CE)

Description: Henry Ford revolutionized the automobile industry with the mass-produced Model T, making cars affordable for the average American family.

History: The assembly line production method used by Ford dramatically reduced the cost of manufacturing, setting the standard for industrial production across various sectors.

43. PENICILLIN (1928 CE)

Description: Discovered by Alexander Fleming, penicillin became the first widely used antibiotic, revolutionizing medicine and saving countless lives from bacterial infections.

History: Penicillin's widespread use during World War II and after led to the development of more antibiotics, which significantly reduced the mortality rates of infectious diseases.

44. THE RADIO (1901 CE)

Description: Guglielmo Marconi successfully transmitted the first radio signal across the Atlantic, proving that long-distance wireless communication was possible.

History: Radio became the first mass communication medium, transforming news, entertainment, and military communication during the 20th century.

45. THE VACUUM CLEANER (1901 CE)

Description: Hubert Cecil Booth invented the first powered vacuum cleaner, which used suction to clean carpets and floors.

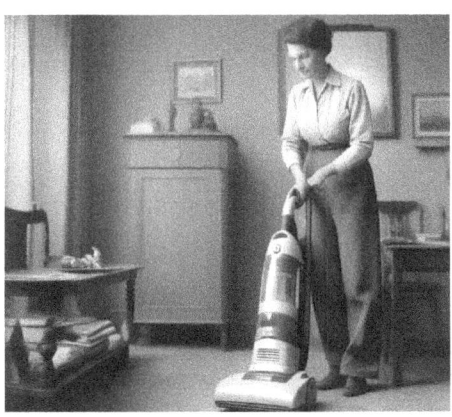

History: The vacuum cleaner became a household staple, significantly reducing the time and effort required for cleaning and improving indoor hygiene.

46. PLASTIC (1907 CE)

Description: Leo Baekeland developed Bakelite, the first fully synthetic plastic, marking the beginning of the modern plastic industry.

History: Bakelite was used in everything from electrical components to household goods. Plastics eventually became one of the most versatile materials of the 20th century, though concerns about pollution have since risen.

47. THE X-RAY MACHINE (1895 CE)

Description: Wilhelm Röntgen discovered X-rays and developed the first X-ray machine, allowing doctors to see inside the human body without surgery.

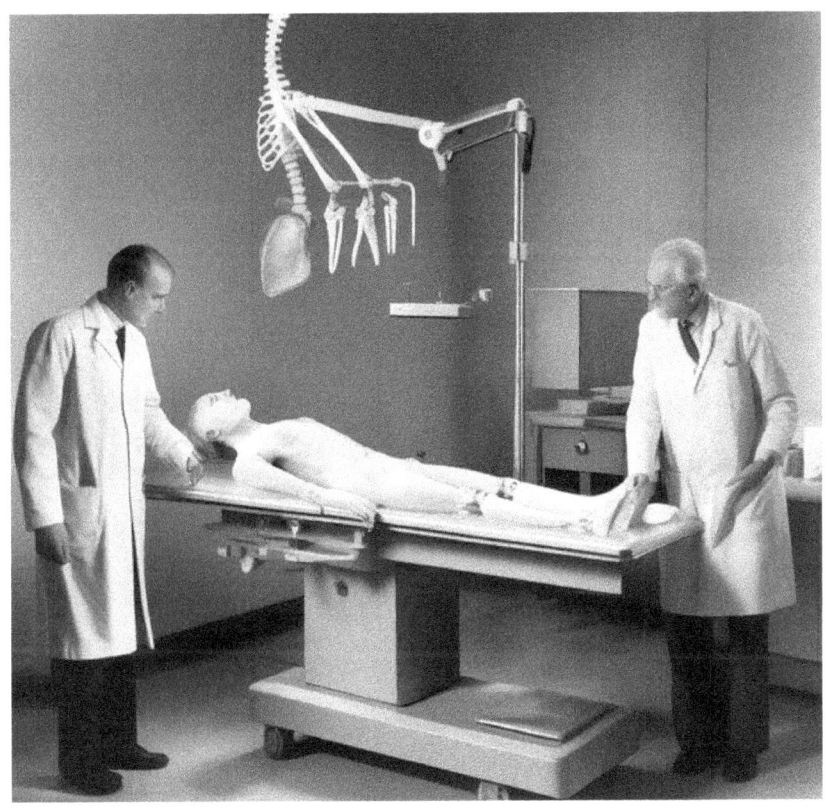

History: X-rays revolutionized medical diagnosis by enabling the non-invasive observation of bones and organs, becoming an indispensable tool in hospitals worldwide.

48. THE HELICOPTER (1936 CE)

Description: Igor Sikorsky developed the first successful mass-produced helicopter, which could take off and land vertically.

History: Helicopters became essential for military operations, rescue missions, and transportation in areas where fixed-wing aircraft could not operate, like dense forests or mountains.

49. THE REFRIGERATOR (1913 CE)

Description: Fred W. Wolf designed one of the first domestic refrigerators, which dramatically improved food storage by keeping perishables fresh for longer periods.

History: The refrigerator transformed the way people stored and consumed food, reducing food waste and making fresh produce available year-round.

50. THE NEON LIGHT (1910 CE)

Description: Georges Claude developed neon lighting, which used electrified neon gas to produce bright, colorful light.

History: Neon lights became a staple of advertising and cityscapes, especially in the mid-20th century, creating iconic visuals in places like Las Vegas and Times Square.

CHAPTER 6: MID-20TH CENTURY INVENTIONS

51. THE JET ENGINE (1930S CE)

Description: Frank Whittle in the UK and Hans von Ohain in Germany developed the jet engine, which allowed for much faster air travel than propeller-driven aircraft.

History: Jet engines transformed both commercial aviation and military airpower, enabling long-distance flights and ushering in the era of the jetliner.

52. TELEVISION (1927 CE)

Description: Philo Farnsworth and John Logie Baird developed the first television systems capable of transmitting moving images, bringing entertainment and news into homes.

History: Television became a dominant form of mass media in the 20th century, shaping culture, politics, and social behavior through its widespread reach.

53. NUCLEAR POWER (1940S CE)

Description: Nuclear fission, first harnessed during World War II, was later used to generate electricity in civilian nuclear power plants.

History: While nuclear power provided a nearly limitless energy source, its development also led to the creation of nuclear weapons, sparking global debates about its risks and benefits.

54. THE COMPUTER (1940S CE)

Description: Early computers, such as the ENIAC, were developed during World War II to calculate artillery trajectories, but they laid the groundwork for the modern digital age.

History: Advances in computing technology during the late 20th century led to the development of personal computers, transforming the way people work, communicate, and access information.

55. THE MICROWAVE OVEN (1945 CE)

Description: Percy Spencer discovered that microwaves could heat food, leading to the invention of the microwave oven, a fast and convenient cooking device.

History: The microwave oven revolutionized food preparation in homes and businesses, allowing for quick reheating and cooking of meals.

56. VELCRO (1941 CE)

Description: Swiss engineer George de Mestral invented Velcro after observing how burrs stuck to his clothing. He mimicked the hook-and-loop structure to create a versatile fastener.

History: Velcro became widely used in fashion, aerospace, and medical devices due to its simplicity and effectiveness, particularly for items needing quick fastening and unfastening.

57. THE TRANSISTOR (1947 CE)

Description: Developed by John Bardeen, William Shockley, and Walter Brattain, the transistor replaced vacuum tubes in electronics, making devices smaller, faster, and more reliable.

History: The transistor revolutionized electronics, leading to the development of modern computers, smartphones, and countless other technologies.

58. THE CREDIT CARD (1950 CE)

Description: Frank McNamara introduced the first credit card, the Diners Club card, which allowed users to charge meals at participating restaurants.

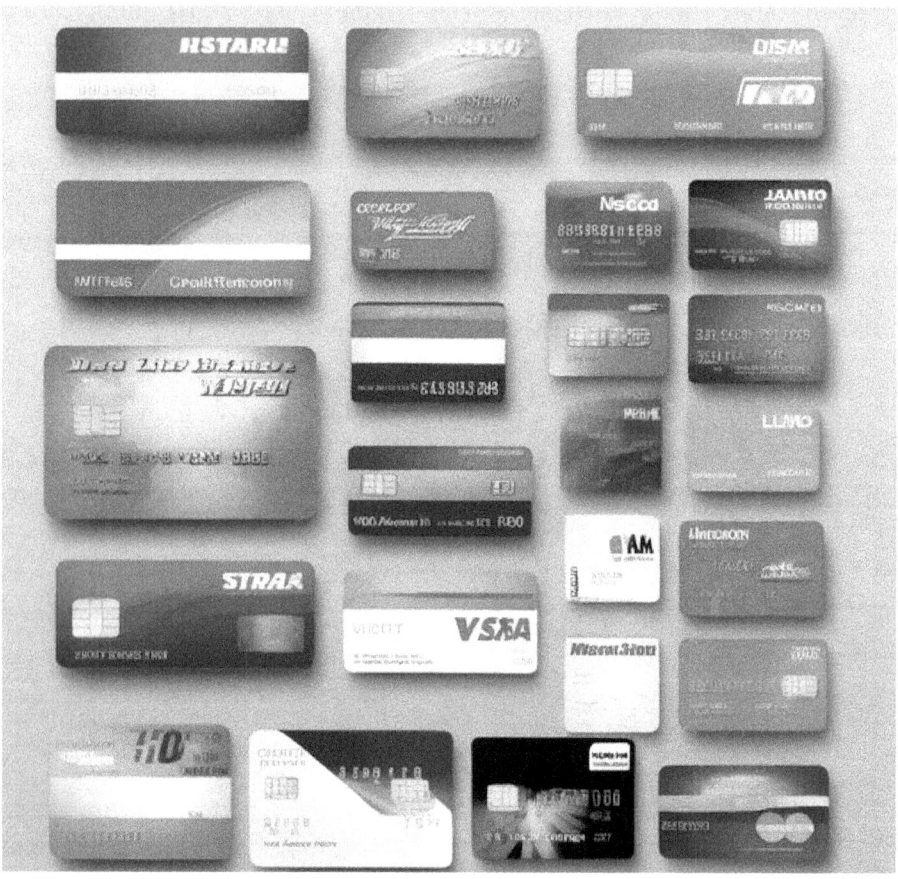

History: Credit cards transformed personal finance, enabling consumers to make purchases on credit and facilitating the rise of the modern consumer economy.

59. THE HOVERCRAFT (1956 CE)

Description: Sir Christopher Cockerell invented the hovercraft, a vehicle that could travel over water, land, and other surfaces by riding on a cushion of air.

History: Hovercrafts became useful in military, rescue, and transportation roles, especially in areas where traditional boats or vehicles struggled to operate.

60. THE POLIO VACCINE (1955 CE)

Description: Jonas Salk developed the first effective polio vaccine, which virtually eliminated the disease in many parts of the world.

History: The polio vaccine was a major public health breakthrough, protecting millions from the debilitating disease and inspiring global vaccination campaigns.

CHAPTER 7: LATE 20TH CENTURY INVENTIONS

61. THE INTERNET (1960S-1990S CE)

Description: Originally developed by the U.S. Department of Defense's ARPANET program, the internet evolved into a global communication network with the advent of the World Wide Web in the 1990s.

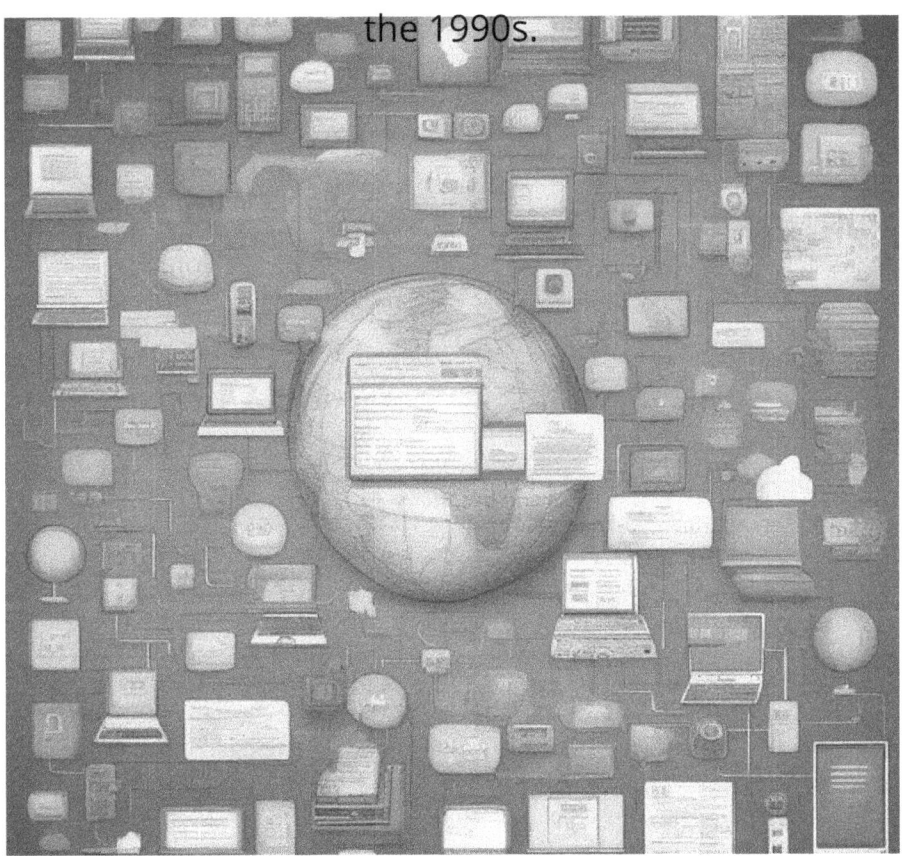

History: The internet revolutionized communication, business, entertainment, and education, connecting billions of people and reshaping industries worldwide.

62. THE PERSONAL COMPUTER (1970S CE)

Description: Early personal computers like the Apple I and the IBM PC made computing accessible to individuals, businesses, and schools.

History: Personal computers transformed the way people worked and learned, leading to the digital revolution that continues to evolve today.

63. THE MOBILE PHONE (1973 CE)

Description: Martin Cooper, a Motorola executive, made the first mobile phone call, paving the way for portable, wireless communication.

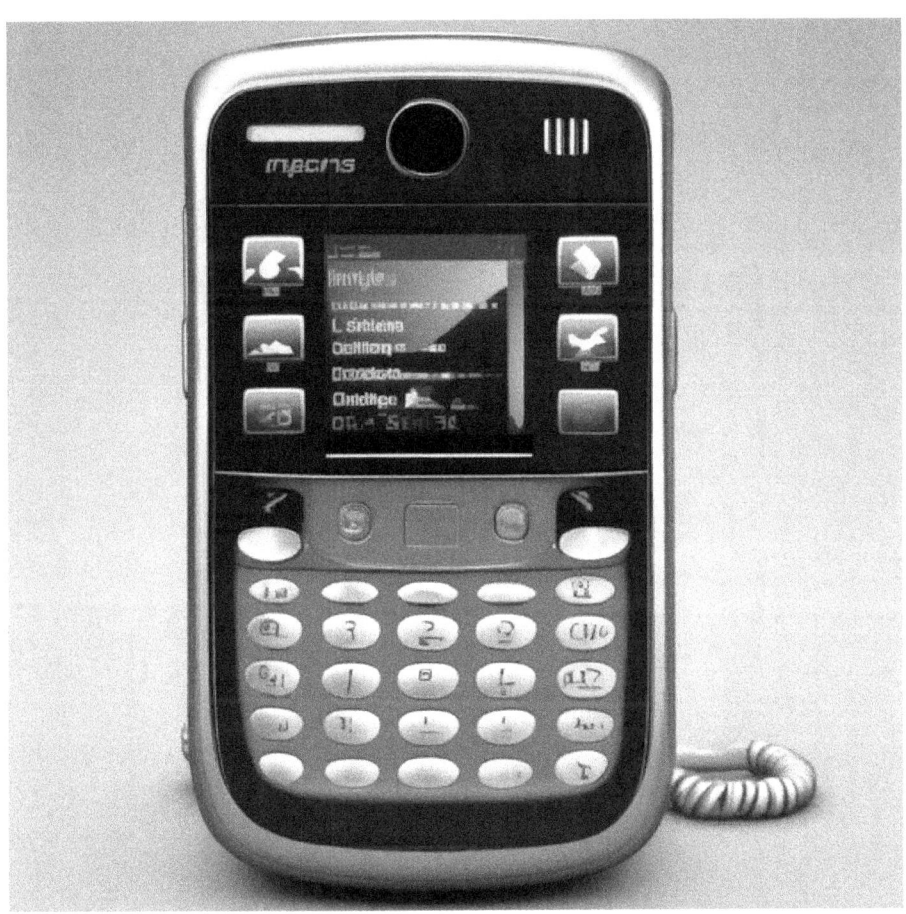

History: The mobile phone evolved from bulky devices to sleek smartphones, transforming global communication and becoming an essential tool for daily life.

64. THE ARTIFICIAL HEART (1982 CE)

Description: Dr. Barney Clark received the first permanent artificial heart implant, developed by a team of engineers and surgeons, giving hope to patients with severe heart disease.

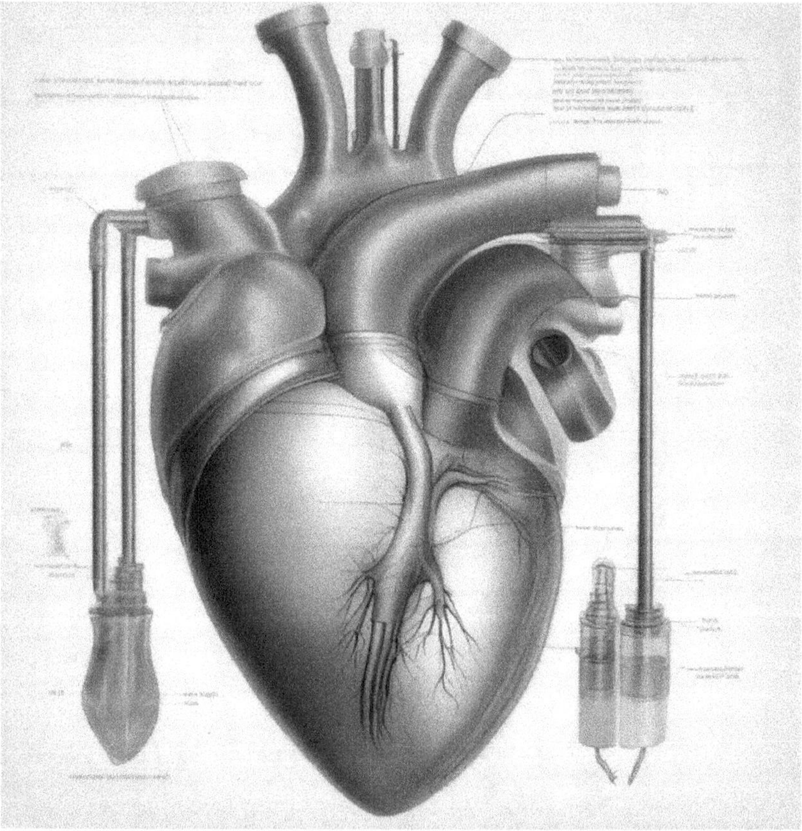

History: While initially used as a temporary measure, artificial hearts have since advanced, extending the lives of patients awaiting transplants.

65. GPS (1973 CE)

Description: The Global Positioning System, developed by the U.S. military, allows users to determine their precise location anywhere on Earth using satellite signals.

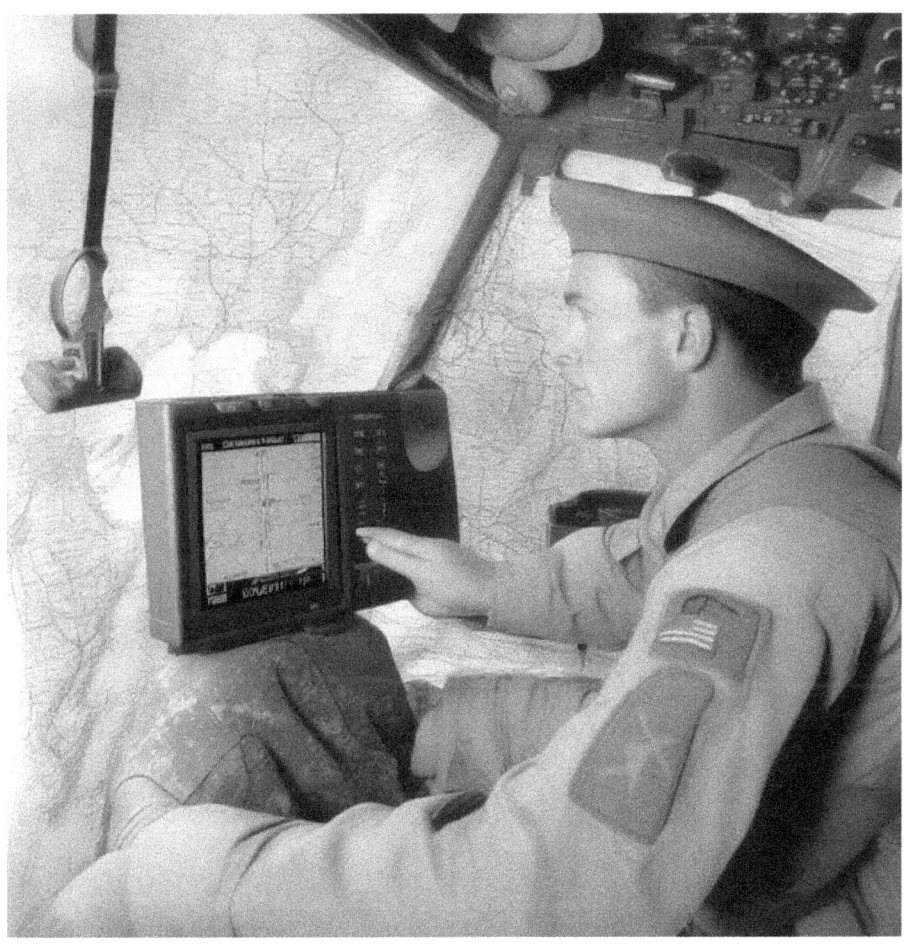

History: Originally for military use, GPS became widely available for civilian use in the 1990s, revolutionizing navigation, mapping, and location-based services.

66. THE SPACE SHUTTLE (1981 CE)

Description: NASA's Space Shuttle program introduced the first reusable spacecraft, capable of transporting astronauts and cargo to space and back.

History: The Space Shuttle was crucial for deploying satellites, building the International Space Station, and conducting scientific research, though it was retired in 2011.

67. DNA FINGERPRINTING (1984 CE)

Description: Sir Alec Jeffreys developed DNA fingerprinting, a technique used to identify individuals based on their unique genetic makeup.

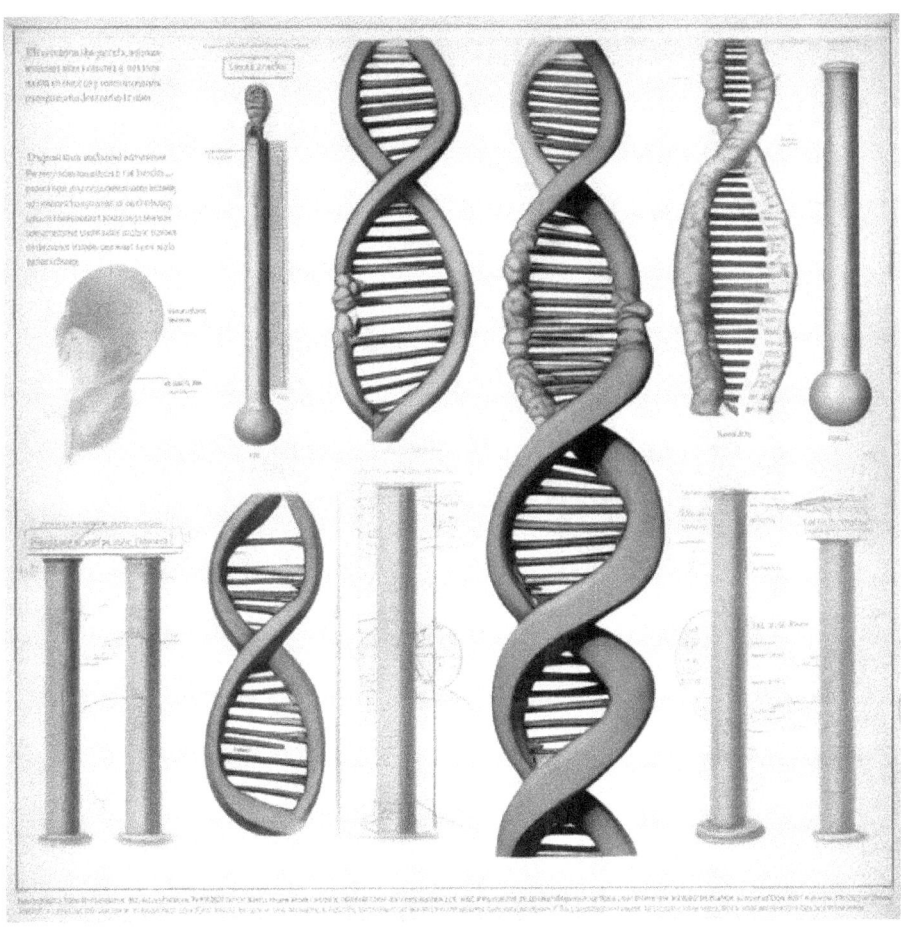

History: DNA fingerprinting revolutionized forensics, paternity testing, and genetic research, becoming a key tool in law enforcement and biology.

68. THE ATM (1967 CE)

Description: John Shepherd-Barron invented the first automated teller machine (ATM), allowing people to withdraw money from their bank accounts 24/7.

History: ATMs revolutionized banking by making it more convenient and accessible for customers, reducing the need for in-person transactions at branches.

69. POST-IT NOTES (1974 CE)

Description: Arthur Fry, a 3M scientist, developed Post-It Notes by combining a weak adhesive with paper, creating a convenient, reusable note that could be stuck to various surfaces.

History: Post-It Notes became an indispensable office tool for jotting down reminders, notes, and ideas, contributing to better organization and productivity.

70. THE MRI SCANNER (1977 CE)

Description: Raymond Damadian developed the first magnetic resonance imaging (MRI) scanner, a medical device that creates detailed images of the body's internal structures.

History: MRI scans revolutionized medical diagnosis by allowing non-invasive imaging of soft tissues, providing detailed views of organs and muscles without the need for X-rays or surgery.

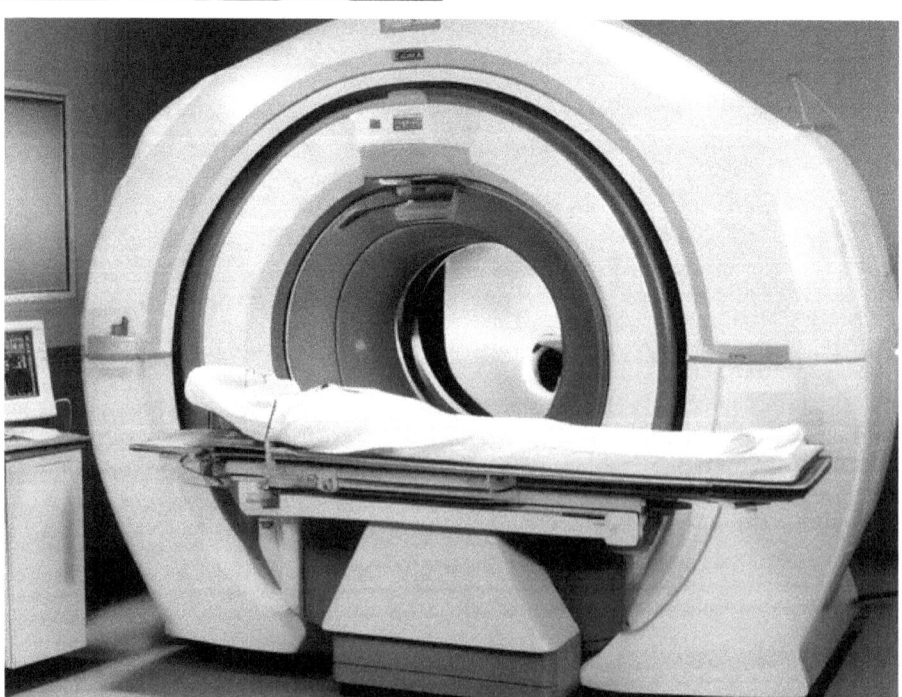

CHAPTER 8: EARLY 21ST CENTURY INVENTIONS

71. THE HYBRID CAR (1997 CE)

Description: Toyota's Prius was the first mass-produced hybrid vehicle, combining a gasoline engine with an electric motor to reduce fuel consumption and emissions.

History: The hybrid car was a breakthrough in environmentally friendly transportation, inspiring other manufacturers to develop fuel-efficient vehicles and leading to the rise of electric cars.

72. SOCIAL MEDIA PLATFORMS (2000S CE)

Description: Platforms like Facebook, Twitter, and Instagram transformed how people communicate, share information, and build communities online.

History: Social media reshaped personal relationships, politics, and business by creating platforms for instant, global communication and content sharing.

73. THE IPHONE (2007 CE)

Description: Developed by Apple, the iPhone combined a phone, a portable media player, a camera, and an internet browser into one device, revolutionizing mobile technology.

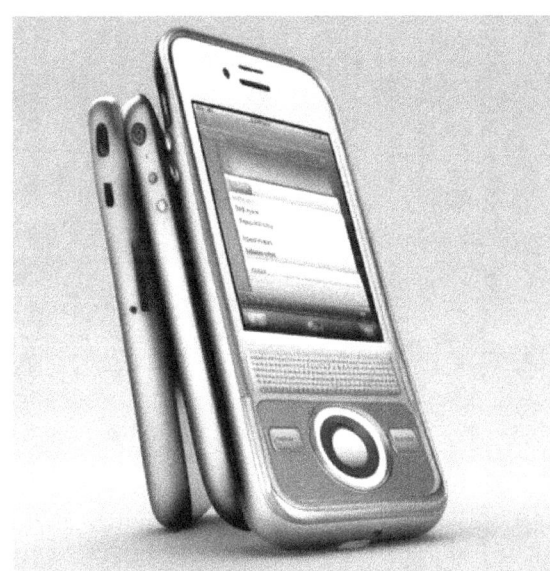

History: The iPhone set the standard for smartphones, becoming a cultural phenomenon and transforming industries like telecommunications, photography, and app development.

74. 3D PRINTING (1980S-2000S CE)

Description: Chuck Hull invented the first 3D printer, a machine that uses additive manufacturing to create objects layer by layer from digital designs.

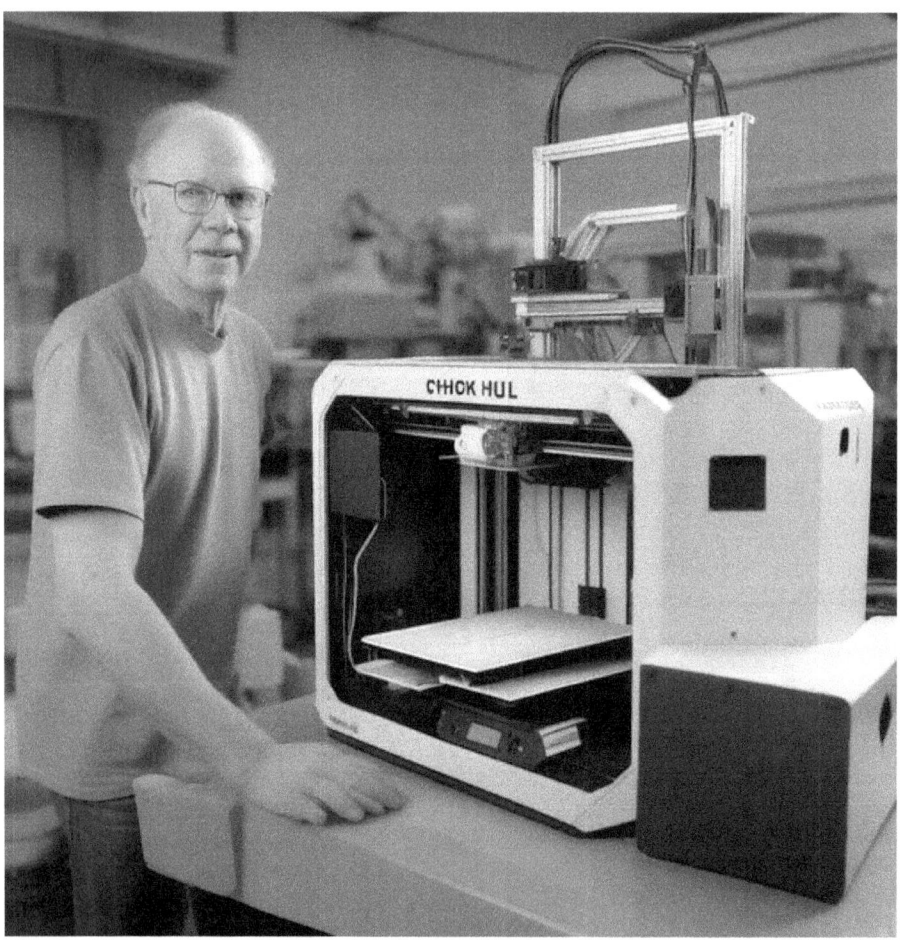

History: 3D printing revolutionized prototyping and manufacturing by allowing rapid production of complex parts, from consumer products to medical implants and aerospace components.

75. BITCOIN (2009 CE)

Description: Bitcoin, the world's first cryptocurrency, was created by an anonymous person or group known as Satoshi Nakamoto. It introduced blockchain technology, allowing decentralized, peer-to-peer transactions.

History: Bitcoin sparked the rise of cryptocurrencies, challenging traditional banking systems and creating a new digital economy, though it remains controversial due to regulatory concerns and volatility.

76. CRISPR GENE EDITING (2012 CE)

Description: CRISPR-Cas9, developed by Jennifer Doudna and Emmanuelle Charpentier, allows scientists to precisely edit genes, offering the potential to treat genetic disorders.

History: CRISPR revolutionized biotechnology, opening up possibilities for gene therapy, agriculture, and even the potential to eliminate diseases like cystic fibrosis and sickle cell anemia.

77. THE ELECTRIC CAR (2008 CE)

Description: Tesla's introduction of the fully electric Roadster demonstrated that electric vehicles could be practical and high-performance, kickstarting the modern electric car revolution.

History: Electric vehicles (EVs) gained mainstream attention as a sustainable alternative to gasoline cars, driven by advancements in battery technology and growing concerns about climate change.

78. CLOUD COMPUTING (2000S CE)

Description: Services like Amazon Web Services (AWS), Google Cloud, and Microsoft Azure allow users to store and access data and applications remotely over the internet.

History: Cloud computing transformed the way businesses and individuals manage data, enabling more flexible, scalable, and cost-effective computing resources.

79. ARTIFICIAL INTELLIGENCE ASSISTANTS (2010S CE)

Description: AI assistants like Apple's Siri, Amazon's Alexa, and Google Assistant use natural language processing and machine learning to help users with tasks like searching the web, managing schedules, and controlling smart devices.

History: AI assistants brought advanced voice recognition and automation into everyday life, making it easier for users to interact with technology and access information hands-free.

80. WEARABLE FITNESS TRACKERS (2010S CE)

Description: Devices like Fitbit and the Apple Watch allow users to track physical activity, heart rate, sleep patterns, and other health metrics.

History: Wearable technology revolutionized personal health monitoring, helping people track fitness goals, manage chronic conditions, and improve overall well-being.

CHAPTER 9: EMERGING INNOVATIONS

81. SELF-DRIVING CARS (2010S CE)

Description: Autonomous vehicles, developed by companies like Tesla, Waymo, and Uber, use sensors, cameras, and AI to navigate roads without human intervention.

History: Self-driving cars promise to reduce traffic accidents and improve transportation efficiency, though regulatory and safety challenges remain.

82. 5G TECHNOLOGY (2020S CE)

Description: The rollout of 5G networks offers faster internet speeds and improved connectivity, enabling innovations in industries like telecommunications, healthcare, and the Internet of Things (IoT).

History: 5G is expected to revolutionize mobile communication and infrastructure by providing ultra-reliable, low-latency connections for smart cities, autonomous vehicles, and more.

83. LAB-GROWN MEAT (2020S CE)

Description: Cultured meat, grown from animal cells in a lab, offers a sustainable alternative to traditional meat production, with the potential to reduce environmental impact and meet global food demands.

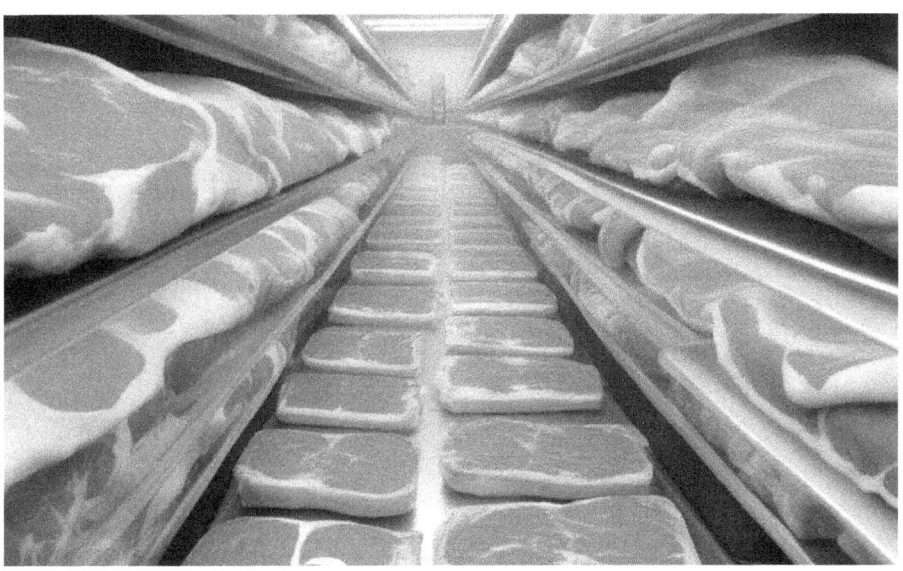

History: Lab-grown meat is in its early stages, but it holds promise for reducing greenhouse gas emissions, land use, and animal suffering associated with conventional agriculture.

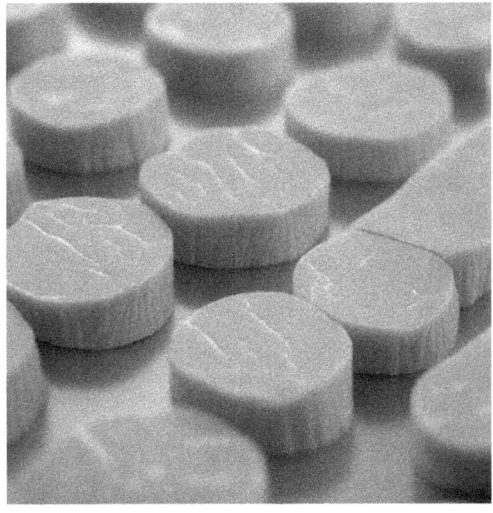

84. QUANTUM COMPUTING (2020S CE)

Description: Quantum computers, developed by companies like IBM, Google, and Rigetti, use the principles of quantum mechanics to process information at unprecedented speeds.

History: Quantum computing could revolutionize fields like cryptography, materials science, and artificial intelligence by solving complex problems that are currently impossible for classical computers.

85. VERTICAL FARMING (2010S CE)

Description: Vertical farms grow crops in stacked layers, often in controlled indoor environments, using techniques like hydroponics and LED lighting to maximize space and reduce water usage.

History: Vertical farming offers a solution to the challenges of urbanization and climate change by producing food locally in urban areas, reducing transportation costs and environmental impact.

86. EXOSKELETONS (2020S CE)

Description: Robotic exoskeletons, developed for medical rehabilitation and industrial applications, augment human strength and mobility, assisting people with disabilities or workers in physically demanding jobs.

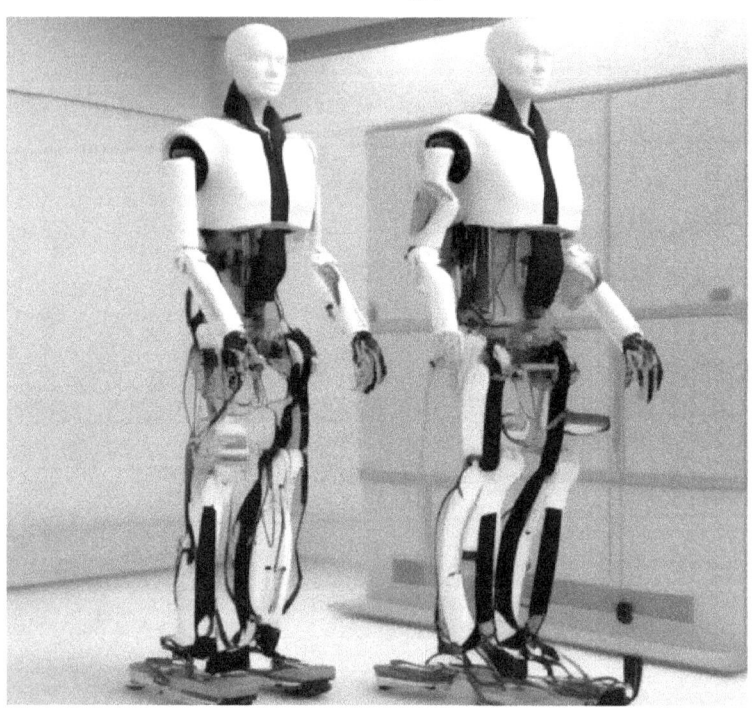

History: Exoskeletons are being used in healthcare to help patients regain mobility after injuries or strokes and in industries like construction and manufacturing to reduce worker fatigue and prevent injuries.

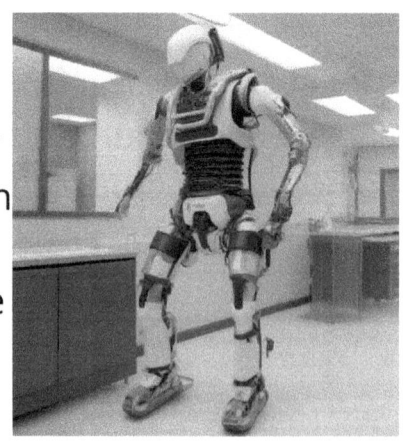

87. HYPERLOOP (2010S CE)

Description: Proposed by Elon Musk, the Hyperloop is a high-speed transportation system that uses low-pressure tubes and magnetic levitation to move pods at speeds of up to 760 mph.

History: Hyperloop technology aims to revolutionize transportation by offering faster, more efficient travel between cities, though it is still in the experimental phase with several companies working on prototypes.

88. AUGMENTED REALITY (2010S CE)

Description: Augmented reality (AR) overlays digital information onto the physical world, allowing users to interact with virtual objects and environments through devices like smartphones or AR glasses.

History: AR is being used in gaming, education, and industry, with applications ranging from immersive learning experiences to remote assistance for complex tasks.

89. HOLOGRAPHIC DISPLAYS (2020S CE)

Description: Holographic display technology creates three-dimensional images that can be viewed without the need for special glasses, offering new ways to visualize and interact with data.

History: Holographic displays have the potential to revolutionize fields like entertainment, education, and communication by offering more immersive and realistic visual experiences.

90. BIODEGRADABLE PLASTICS (2020S CE)

Description: Advances in materials science are leading to the development of eco-friendly plastics that break down in natural environments, reducing pollution and waste.

History: Biodegradable plastics are a promising solution to the global plastic waste crisis, offering alternatives for packaging, consumer products, and medical applications.

CHAPTER 10: FUTURE INVENTIONS
91. ARTIFICIAL GENERAL INTELLIGENCE (AGI)

Description: AGI refers to machines with the ability to perform any intellectual task that a human can do, potentially revolutionizing industries by automating complex problem-solving.

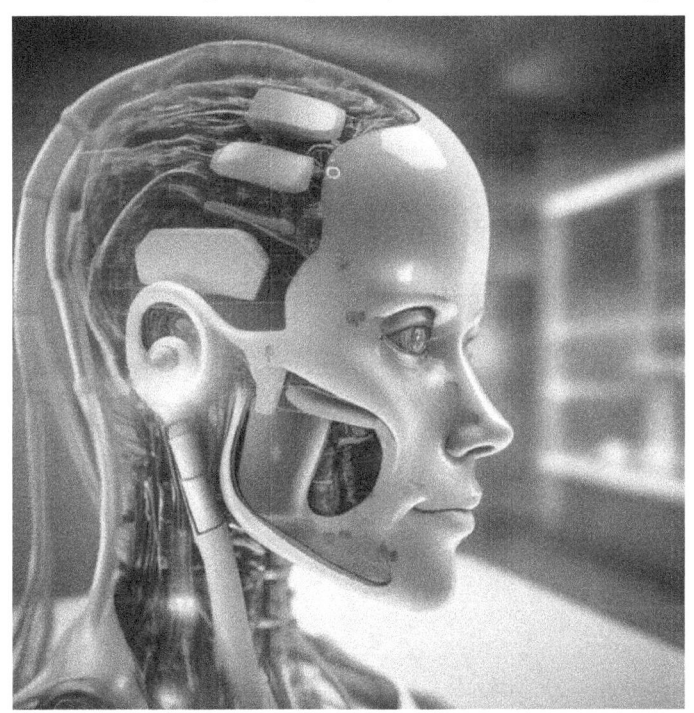

History: While current AI systems are specialized, AGI remains a long-term goal in the field of artificial intelligence, with the potential to transform everything from healthcare to scientific research.

92. FUSION ENERGY

Description: Fusion power aims to replicate the energy-producing process of the sun by fusing atomic nuclei, offering a nearly limitless, clean energy source.

History: Despite decades of research, practical fusion energy remains elusive, but advancements in experimental reactors like ITER offer hope for a future where fusion replaces fossil fuels and fission reactors.

93. SMART CITIES

Description: Smart cities use a network of sensors, AI, and data analytics to manage resources, infrastructure, and services more efficiently, improving quality of life for residents.

History: Smart city projects are being implemented in places like Singapore, Barcelona, and Dubai, where technology is used to manage traffic, energy consumption, and waste, making cities more sustainable and livable.

94. BRAIN-COMPUTER INTERFACES (BCIS)

Description: BCIs allow for direct communication between the brain and external devices, offering potential breakthroughs in medicine, communication, and even human augmentation.

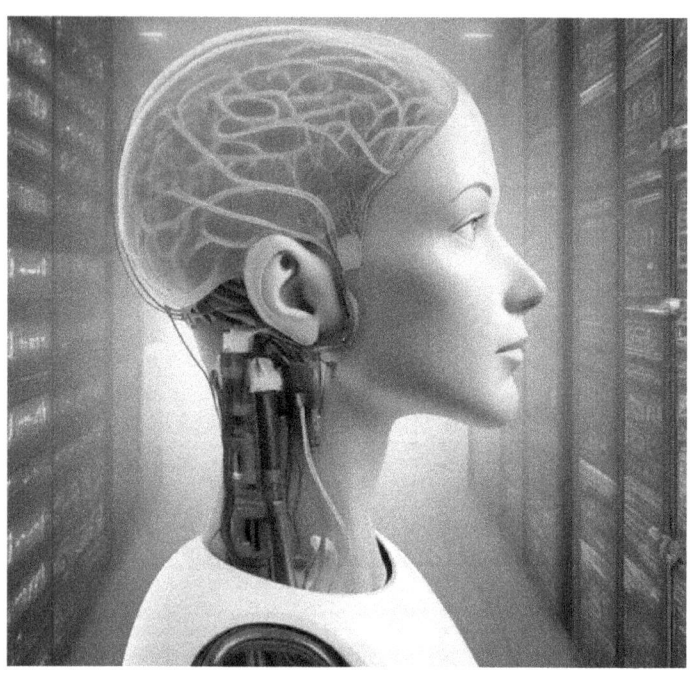

History: Companies like Neuralink are developing BCIs to help people with neurological disorders, such as paralysis or ALS, control devices with their thoughts, opening the door to future applications in learning, gaming, and more.

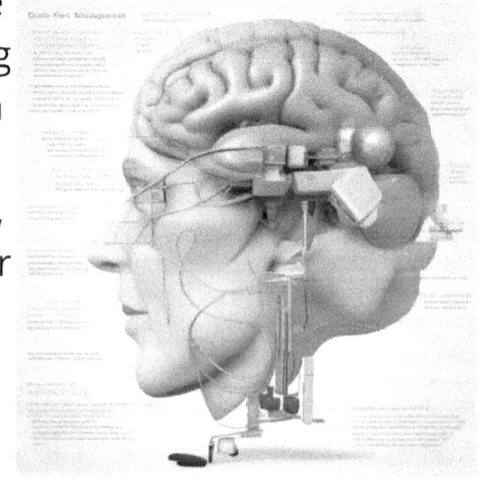

95. TELEPORTATION

Description: Though still theoretical, teleportation based on quantum entanglement allows for the instantaneous transfer of information, and one day, it could make physical teleportation a reality.

History: Scientists have successfully demonstrated quantum teleportation of particles, sparking interest in future applications in communication, computing, and even transportation.

96. SPACE TOURISM

Description: Companies like SpaceX, Blue Origin, and Virgin Galactic are developing spacecraft to make space travel accessible to private citizens, paving the way for a new industry.

History: Space tourism is in its early stages, with suborbital flights already taking place and plans for orbital and even lunar tourism on the horizon, offering a new frontier for exploration and entertainment.

97. ARTIFICIAL ORGANS

Description: Advances in biotechnology and 3D printing are leading to the development of artificial organs, which could one day eliminate the need for organ transplants.

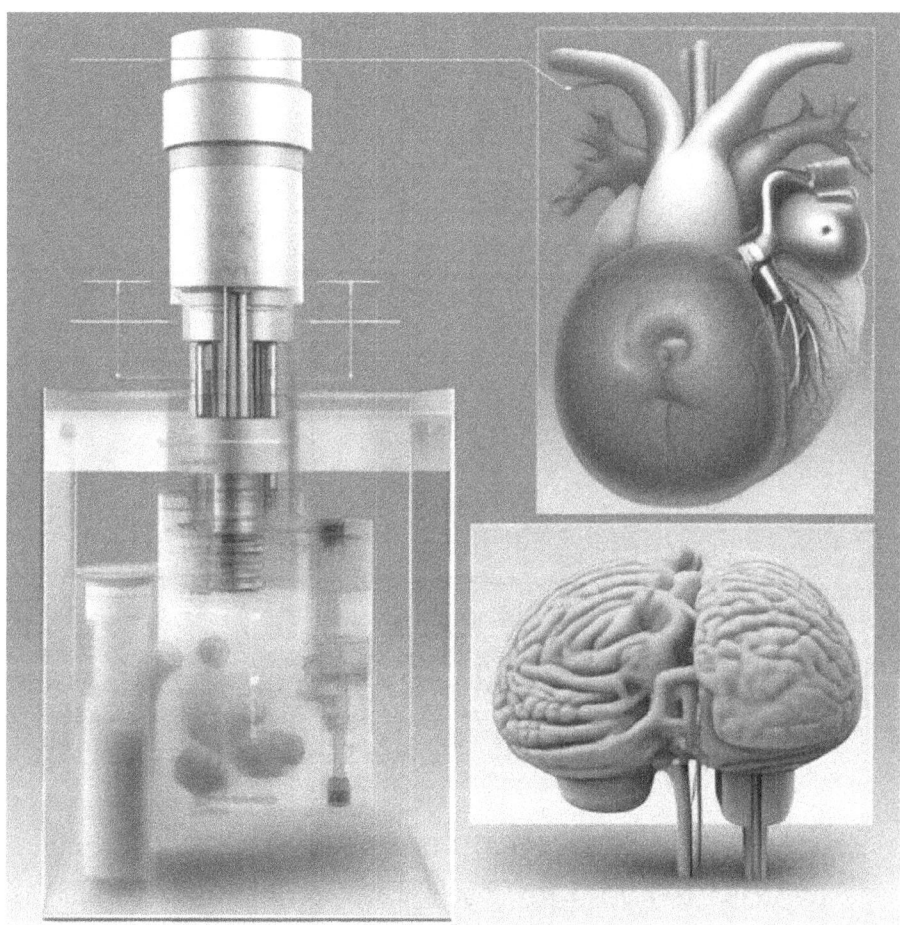

History: Scientists have already 3D printed functional tissues like skin and blood vessels, and the ultimate goal is to create fully functional hearts, kidneys, and other organs to save lives and reduce transplant waiting times.

98. NANOMEDICINE

Description: Nanotechnology in medicine involves the use of tiny, nanoscale materials and devices to deliver drugs, repair cells, and treat diseases at the molecular level.

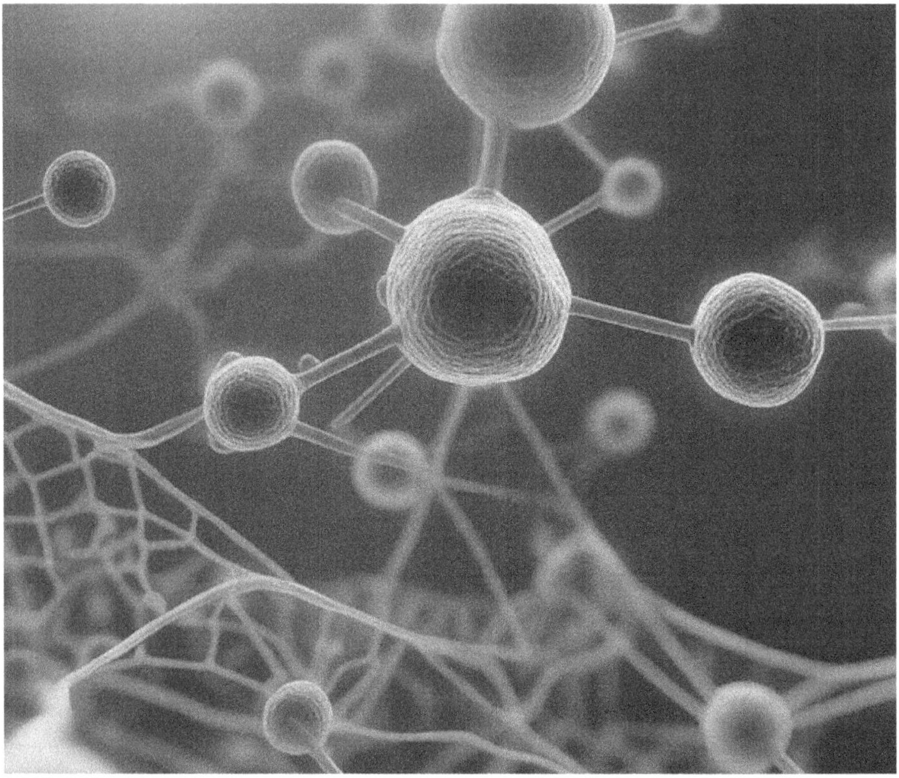

History: Nanomedicine is a rapidly growing field with potential applications in cancer treatment, gene therapy, and diagnostics, offering targeted, precise treatments that minimize side effects.

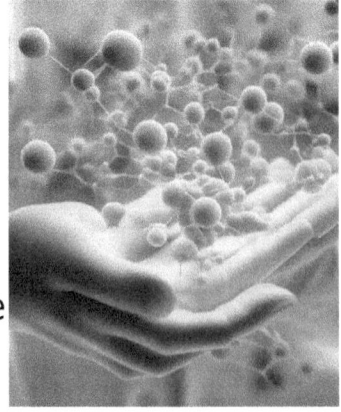

99. FLYING CARS

Description: Flying cars, long a dream of science fiction, are being developed by companies like Terrafugia and PAL-V, with the goal of providing a new form of personal transportation.

History: Flying cars face numerous technical and regulatory challenges, but advances in aviation and electric propulsion bring the possibility of airborne commuting closer to reality.

100. INTERSTELLAR TRAVEL

Description: Interstellar travel would allow humans to explore other star systems, potentially using advanced technologies like warp drives or generation ships for long-duration space missions.

History: Though still in the realm of theoretical physics, interstellar travel is a topic of interest for scientists and space agencies, with projects like Breakthrough Starshot exploring the possibility of sending probes to nearby stars like Alpha Centauri.

CONCLUSION:

As we've journeyed through the remarkable history of human innovation, from the earliest tools to the technologies shaping our future, we hope this book has provided you with a deeper appreciation for the inventors and ideas that have transformed our world. These 100 inventions are more than just objects or technologies—they are reflections of human curiosity, creativity, and determination.

If you enjoyed The Evolution of Innovation: 100 Inventions That Changed the World and found value in the stories behind these groundbreaking creations, we would love to hear your thoughts! Your feedback is invaluable and helps us improve, inspire, and continue sharing the incredible tales of human achievement.

Please consider leaving a review—your input not only supports the book but also helps future readers discover the fascinating world of innovation. Thank you for joining us on this exploration of human ingenuity, and we hope this journey sparks your own curiosity and creativity!